WHAT MAKES LIGHTNING AND THUNDER?

By Marie Roesser

Gareth Stevens
PUBLISHING

Please visit our website, www.garethstevens.com. For a free color catalog of all our high-quality books, call toll free 1-800-542-2595 or fax 1-877-542-2596.

Cataloging-in-Publication Data

Names: Roesser, Marie.
Title: What makes lightning and thunder? / Marie Roesser.
Description: New York : Gareth Stevens Publishing, 2021. | Series: Everyday mysteries | Includes glossary and index.
Identifiers: ISBN 9781538256473 (pbk.) | ISBN 9781538256497 (library bound) | ISBN 9781538256480 (6 pack)
Subjects: LCSH: Lightning–Juvenile literature. | Thunder–Juvenile literature. | Thunderstorms–Juvenile literature.
Classification: LCC QC966.5 R64 2021 | DDC 551.56′32–dc23

Published in 2021 by
Gareth Stevens Publishing
111 East 14th Street, Suite 349
New York, NY 10003

Copyright © 2021 Gareth Stevens Publishing

Editor: Therese Shea

Photo credits: Cover, p. 1 swa182/Shutterstock.com; pp. 3–24 (background) Natutik/Shutterstock.com; p. 5 Vasin Lee/Shutterstock.com; p. 7 Pictureguy/Shutterstock.com; p. 9 Dn Br/Shutterstock.com; p. 11 valdezrl/Shutterstock.com; pp. 13, 17 John D Sirlin/Shutterstock.com; p. 15 OSweetNature/Shutterstock.com; p. 19 Fouad A. Saad/Shutterstock.com; p. 21 Designua/Shutterstock.com.

Printed in the United States of America

CPSIA compliance information: Batch #CS20GS: For further information contact Gareth Stevens, New York, New York at 1-800-542-2595.

Find us on

CONTENTS

Boldface words appear in the glossary.

Bright Lights, Big Noises

Years ago, people made up all kinds of stories to explain flashes of light and booming sounds coming from the sky. In the 1700s, scientists discovered what's happening in the air to cause lightning and thunder. Read on to find out!

It's Electric!

Lightning is **electricity**! Here's why it forms. Winds in rain clouds push around bits of water and ice. As the **particles** crash, they gain charges, or amounts of electricity. There are two kinds of electrical charges: positive charges and negative charges.

Negative charges gather in the middle or bottom of the rain cloud. Positive charges gather at the top. As the rain cloud grows larger, more charges gather. Electrical charges that are alike repel, or push each other away. Different charges **attract** each other.

like charges repel

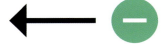 

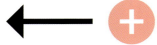 

different charges attract

Often, the ground has a weak negative charge. However, a rain cloud's negative charge can repel negative charges on the ground. Then, positive charges gather below the cloud. The negative and positive charges are attracted to each other.

The electrical charges need to be very strong to meet through the air. Then, the cloud sends a negatively charged electric **current** toward the ground. A positively charged current comes up from the ground to meet it. We see these currents as lightning!

13

Cracks and Rumbles

Air is made up of gases. Gases **expand** when they heat up. They **contract** when they cool. This is why thunder happens. The electricity in lightning heats the air around it to high **temperatures**. The temperatures can be hotter than the sun's **surface**!

cooled gas contracts

heated gas expands

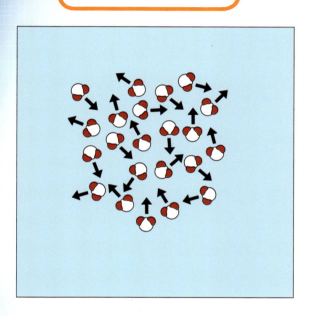

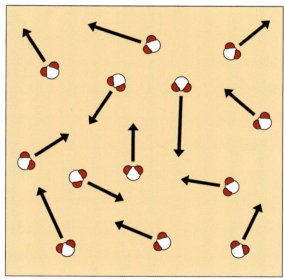

 = gas particle

The air around lightning expands really fast. That makes the air beyond this heated air move too. After the lightning, the air cools quickly and contracts. This causes the loud, sharp sound we hear as a crack of thunder.

As the air continues to move, you might hear quieter rumbles of thunder. You often hear thunder after you see lightning. Light travels through the air faster than sound. If you hear and see them at the same time, the lightning is close!

LIGHT TRAVELS FASTER THAN SOUND

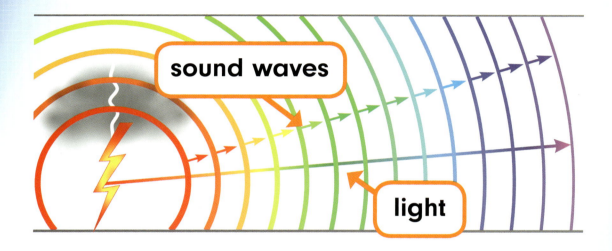

sound waves

light

speed of light = **186,282 miles per second**
(299,792 km/s)

speed of sound = **1,087 feet per second**
(331 m/s)

19

Lightning Safety

Lightning also can happen within a cloud and between clouds. It can reach from a cloud to a charged object on the ground too. Lightning can harm and even kill people. That's why it's important to stay inside during storms.

HOW LIGHTNING AND THUNDER FORM

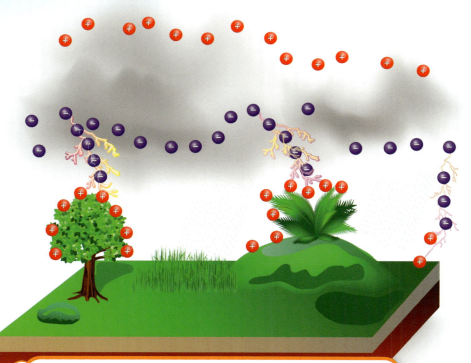

1. Positive and negative particles separate in rain clouds.
2. Positive particles gather below the cloud.
3. Negative particles attract positive particles.
4. Electric currents connect the charges and make lightning.
5. Air movement around lightning causes thunder.

21

GLOSSARY

attract: to draw nearer

contract: to get smaller and tighter

current: a flow of electricity resulting from the movement of particles such as electrons

electricity: a form of energy, or power, created by charged particles

expand: to get larger and looser

particle: a very small piece of something

surface: the top layer of something

temperature: how hot or cold something is

22

FOR MORE INFORMATION

BOOKS

Lawrence, Ellen. *What Is Lightning?* New York, NY: Bearport Publishing, 2016.

Rivera, Andrea. *Thunder & Lightning*. Minneapolis, MN: Abdo Zoom, 2017.

WEBSITES

Electricity for Kids
www.ducksters.com/science/electricity_101.php
Learn what makes an electrical charge positive or negative.

Lightning
kids.nationalgeographic.com/explore/science/lightning-/
Find out how many lightning strikes are happening right now!

Lightning Myths and Facts
www.weather.gov/safety/lightning-myths
Learn many more amazing lightning facts.

INDEX